W9-DIY-888

The Future of Renewable Energy

What Is the Future of Alternative Energy Cars?

Lydia Bjornlund

ReferencePoint Press®

San Diego, CA

© 2014 ReferencePoint Press, Inc.
Printed in the United States

For more information, contact:
ReferencePoint Press, Inc.
PO Box 27779
San Diego, CA 92198
www.ReferencePointPress.com

Cover: Thinkstock Images
Maury Aaseng: 15, 21, 28, 34, 41, 46, 53, 61
© Kai Pfaffenbach/Reuters/Corbis: 9

LIBRARY OF CONGRESS CATALOGING-IN-PUBLICATION DATA

Bjornlund, Lydia, 1961–
  What is the future of alternative energy cars? / by Lydia Bjornlund.
    pages cm. — (Future of renewable energy series)
  Includes bibliographical references and index.
    ISBN 978-1-60152-610-6 (hardback)—ISBN 1-60152-610-5 (hardback) 1. Alternative fuel vehicles.
  2. Automobiles—Fuel consumption. 3. Automobiles—Technological innovations. I. Title.
  TL216.5.B56 2014
  388.3'4--dc23
                                                                    2013036244

Contents

Foreword

What are the long-term prospects for renewable energy?

In his 2011 State of the Union address, President Barack Obama set an ambitious goal for the United States: to generate 80 percent of its electricity from clean energy sources, including renewables such as wind, solar, biomass, and hydropower, by 2035. The president reaffirmed this goal in the March 2011 White House report *Blueprint for a Secure Energy Future*. The report emphasizes the president's view that continued advances in renewable energy are an essential piece of America's energy future. "Beyond our efforts to reduce our dependence on oil," the report states, "we must focus on expanding cleaner sources of electricity, including renewables like wind and solar, as well as clean coal, natural gas, and nuclear power—keeping America on the cutting edge of clean energy technology so that we can build a 21st century clean energy economy and win the future."

Obama's vision of America's energy future is not shared by all. Benjamin Zycher, a visiting scholar at the American Enterprise Institute, a conservative think tank, contends that policies aimed at shifting from conventional to renewable energy sources demonstrate a "disconnect between the rhetoric and the reality." In *Renewable Electricity Generation: Economic Analysis and Outlook* Zycher writes that renewables have inherent limitations that can be overcome only at a very high cost. He states: "Renewable electricity has only a small share of the market, and ongoing developments in the market for competitive fuels . . . make it likely that renewable electricity will continue to face severe constraints in terms of competitiveness for many years to come."

Is Obama's goal of 80 percent clean electricity by 2035 realistic? Expert opinions can be found on both sides of this question and on all of the other issues relating to the debate about what lies ahead for renewable energy. Driven by this reality, *The Future of Renewable Energy* series critically examines the long-term prospects for renewable energy by delving into the topics and opinions that dominate and inform renewable energy policy and debate. The series covers renewables such as solar, wind, biofuels, hydrogen, and hydropower and explores the issues of cost and affordability, impact on the environment, viability as a replacement for fossil fuels, and what role—if any—government should play in renewable energy development. Pointed questions (such as "Can Solar Power Ever Replace Fossil Fuels?" or "Should Government Play a Role in Developing Biofuels?") frame the discussion and support inquiry-based learning. The pro/con format of the series encourages critical analysis of the topics and opinions that shape the debate. Discussion in each book is supported by current and relevant facts and illustrations, quotes from experts, and real-world examples and anecdotes. Additionally, all titles include a list of useful facts, organizations to contact for further information, and other helpful sources for further reading and research.

Visions of the Future: Alternative Energy Cars

In 2012 the Tesla Model S, an all-electric sedan, hit the US market. There were already electric vehicles on the road, but this one was different. Unlike the tiny, boxy, odd-looking electric cars of earlier generations, this was a sleek, stylish vehicle that spelled luxury. "The Model S epitomizes efficiency, embodying the grace and performance of a world-class athlete. Its sculpted form expresses a constant state of speed and motion,"[1] says Franz von Holzhausen, Tesla's chief designer. But the Model S did not just look good, it outperformed the other luxury vehicles in its class, offering superior acceleration and handling.

The fact that the Model S was an electric vehicle got it immediate attention, but automobile enthusiasts soon looked beyond this fact to compare it to conventional vehicles. It won coveted awards from automobile magazines and international design associations. *Consumer Reports* gave it the highest rating ever—99 out of 100 points. And it won the hearts of consumers. In the first quarter of 2013 the Model S outsold similarly priced BMW and Mercedes models and helped Tesla turn a profit. For many, the Model S was a sign of the future—a future in which the gasoline-powered car with its internal combustion engine was a thing of the past. "The Tesla Model S is just the beginning of a journey to change consumer perceptions of electrically powered personal commuting," says Roger Swales, the head of the 2013 International Design Excellence Awards (IDEA) jury of the Industrial Design Society of America (IDSA). "By creating a desirable and viable alternative to gasoline-powered transportation, Tesla has proven that being environmentally responsible does not have to limit the aspiration for desirable product solutions."[2]

Electric-Drive Technologies

The number of cars on the road is skyrocketing. The price of gasoline is escalating. The world's dependence on oil makes it vulnerable to supply shortages and the whims of oil-exporting countries. For these reasons and more, people are looking for alternative energy sources to power their vehicles.

Among the most promising new alternative energy sources is electricity. Electric-drive technologies are a broad category that encompasses battery electric, hybrid electric, plug-in hybrids, and fuel cell vehicles. In a conventional vehicle, the wheels are turned by a mechanical drive that is powered by an internal combustion engine; in an electric-drive vehicle, the wheels are turned by one or more electric motors.

Essentially, there are four types of electric-drive vehicles. Hybrids, or hybrid-electric vehicles (HEVs), combine an internal combustion engine with an electric motor and battery. The battery is recharged during braking and coasting and then used for acceleration or other functions. The extra power of the battery allows the internal combustion engine to be smaller and reduces fuel consumption and tailpipe emissions. Plug-in hybrids are modified hybrids that can be charged for short-range travel on battery power alone. The gas engine kicks in on longer trips, when the battery reaches the end of its range. Fully electric cars do not have an engine at all—instead they run entirely on battery power that is recharged from an electrical outlet. Finally, the fourth type of electric-drive vehicle uses hydrogen gas and includes a fuel cell to convert the hydrogen to energy. Fuel cell technologies have been lauded as the future for many years but have lagged behind other technologies for electric vehicles.

The Electric Car Past and Present

The electric car is nothing new. In fact, it dates back to the birth of the automobile—or even before. In the 1820s and 1830s several inventors developed small-scale model cars powered by an electric motor. The first successful full-scale, drivable model was created in the 1890s, and at the turn of the century there were a number of options to choose from. Early automobile enthusiasts appreciated the advantages that electric cars had over those powered by the internal combustion engine. Electric cars of-fered a smoother, quieter ride; did not smell like gasoline; and did not require changing gears. In 1900, of the 4,192 cars produced in the United States, 28 percent were powered by electricity.

As people began to travel longer distances in their cars, however, the limitations of battery-powered vehicles became increasingly evident, and people began to move to gasoline-powered cars. The electric starter, which provided an alternative to the hand-crank needed to get gasoline engines to start running, meant the end for this first generation of elec-tric vehicles. For the next several decades, cars powered by gasoline or diesel dominated the roads around the world.

But many people continued to wonder whether there was not a bet-ter alternative to the dirty, polluting internal combustion engine. Begin-ning in 1966 Congress introduced a series of bills recommending the use of electric vehicles as a means of reducing air pollution. As concerns about the soaring price of oil grew in the 1970s, so did interest in alter-natively powered vehicles.

In 1990 the California Air Resources Board (CARB) instituted a Zero Emission Vehicle mandate that dictated that 2 percent of vehicles sold in the state of California had to emit zero pollutants, with that number growing to 10 percent by 2003. California represents a huge market—larger than Canada. To capture this market, automobile manufacturers needed to comply with CARB's mandate. Automakers responded with a number of electric models, but the response among consumers was luke-warm. By 2003—the year in which CARB envisioned one in every ten cars to have zero emissions—the generation of electric vehicles inspired by this vision was a thing of the past. The automakers that had offered leases on their electric car models recalled the cars. Many of the cars ended up being crushed so the scrap metal could be melted down and reused.

How They Work

This generation of electric cars worked the same way as those of the pre-vious century. As the name indicates, electric vehicles are propelled by one or more electric motors powered by rechargeable battery packs. The

Some experts say the Tesla Model S car (pictured with an electric vehicle charging station at a 2013 automotive show) represents a future without traditional gasoline-powered cars. Others have their doubts about the long-term viability of alternative energy cars.

main obstacle to the electric vehicle is that when the battery runs out, it has to be recharged. The physics are basic: The larger the battery, the farther it can go without needing a charge. Unfortunately, providing a range of more than one hundred miles or so would require a battery pack that would be too large and heavy for an automobile—a problem that some of the world's best scientists are working to resolve.

In the meantime, engineers came up with an alternative solution: the hybrid electric vehicle. Hybrids were conceived of as a way to address the limitations of batteries by using a combination of gasoline power and electric power. An onboard generator powered by the internal combustion engine is used to charge the battery, giving the vehicle a longer range. Some hybrid vehicles—so-called plug-in hybrids—can charge the battery through an outside source, but most of the popular models on the road today use batteries that are charged by an in-vehicle charging system.

The advantage of a hybrid is that it can use a smaller internal combustion engine because it does not need to achieve peak power from this engine. When additional fuel is needed, to accelerate or climb hills, for example, the auxiliary power comes from a battery that powers an electric motor. In some vehicles the motor powers the vehicle traveling at slow speeds, when the internal combustion engine is least efficient. As in conventional cars, auxiliary functions such as lights and radio usually draw power from a separate lead-acid battery.

New technologies are enhancing the benefits of alternative power sources. For instance, regenerative braking, which was introduced by Toyota in 2000, allows the energy released during coasting or stopping to recharge the storage battery. In regenerative braking, the electric motor applies resistance to the drivetrain, causing the wheels to slow down. In return, the energy from the wheels turns the motor, which functions as a generator, converting energy normally wasted during coasting and braking into electricity, which is stored in a battery until needed by the electric motor.

Some traditional hybrids also are integrating stop-start technologies, which turn off the internal combustion engine when the car is stopped, saving both fuel and emissions. When the driver steps on the gas, the battery propels the car forward and restarts the engine. The stop-start technology is also being used in otherwise conventional automobiles, creating what is sometimes called a micro-hybrid. Stop-start vehicles require more robust batteries and a more complex starter system than is found in a traditional internal combustion vehicle, but the relatively low cost and easy integration of stop-start technologies make them a target for automakers seeking ways to improve gas mileage on traditional and hybrid vehicles.

Hydrogen Fuel Cells–Powered Vehicles

Electric cars are here today, but many experts believe that hydrogen fuel cells will prove to be the technology for tomorrow. Fuel cells are highly technical, but they work like regular batteries, except they never wear out and do not need to be recharged. As long as the cell receives fuel, it will generate electricity. Fuel cell vehicles powered by pure hydrogen emit no harmful air pollutants and can offer fuel equivalent of roughly twice that of gasoline.

The challenge has been to find a safe and effective means of refueling with hydrogen. Some options include a hydrogen gas storage tank in which the gas is stored at extremely high pressure. Another would use hydrogen refueling stations for the fuel cells. Although significant investment has been made into researching the fuel cell technology, engineers have yet to overcome the significant obstacles to making hydrogen-powered vehicles a viable option.

Experts disagree about the future of transportation. Some say that conventional cars will continue to dominate the landscape, but others believe that fuel-cell and/or plug-in electric-drive vehicles will compete with conventional cars within a decade or two. Engineers and designers have made great strides in improving the technology; the number of alternatively powered cars on the roads of tomorrow will depend primarily on how fast additional improvements can be made to give consumers what they want in terms of price, performance, reliability, and range.

Chapter One

Are Alternative Energy Cars Affordable?

Alternative Energy Cars Are Too Expensive

Alternative energy vehicles have proved to be expensive not only to produce but to own. Although they are often touted as a means to save on gasoline, the expense of maintenance and electricity—not to mention the initial costs of ownership—continue to put them out of reach for most drivers.

The Debate

Alternative Energy Cars Are Increasingly Affordable

The costs of alternative energy vehicles have already come down considerably and, in some markets, are already priced competitively, particularly when factoring in the considerable savings in fuel and maintenance costs. As sales increase, economies of scale will bring down production costs.

Alternative Energy Cars Are Too Expensive

"Alternative powertrains face an array of challenges as they attempt to gain widespread acceptance in the market. It is the financial issues that most often resonate with consumers, whether it is the higher price of the vehicle itself, the cost to fuel or charge the vehicle, or the fear of higher maintenance costs. The bottom line is that most consumers want to be green, but not if there is a significant personal cost to them."

—Mike VanNieuwkuyk, executive director of global vehicle research at the marketing information services firm J.D. Power and Associates.

J.D. Power and Associates Reports, *Despite Rising Fuel Prices, the Outlook for Green Vehicles Remains Limited for the Foreseeable Future*, press release, April 27, 2011.

Alternative energy cars cost too much to produce. Unlike vehicles that run on other alternative fuels, such as ethanol or biodiesel, electric and hydrogen fuel cell vehicles require entirely new engineering. This increases the research and development (R&D) cycle and the associated costs.

Compounding the R&D costs is the high cost of the batteries, which sometimes amount to half a vehicle's production costs. Traditional lead-acid batteries cannot achieve the desired range, so automakers have turned to batteries with more expensive base materials, such as lithium, nickel, and zinc. It is time consuming—and thus expensive—to find the best way to craft these materials into functioning modules for electricity storage. While the cost of the batteries varies greatly depending on the size and type, the standard battery for an electric vehicle costs at least

$10,000 and sometimes much more. "The big challenge is reducing the battery price," explains Siddiq Khan, the coauthor of a 2013 report on plug-in electric vehicles. "It constitutes almost a third of the cost of an electric vehicle."[3]

The cost of the battery contributes to the higher production cost of hybrids as well. The only way manufacturers can recoup the added cost is by increasing the price for consumers, but the resulting cost differential has made hybrids and electric cars a tough sell. In 2010 the sale of all hybrids amounted to just 2.4 percent of the total vehicle sales. Sales of the Chevrolet Volt, a plug-in hybrid that entered the market in 2011, have been consistently under projections; the car has done so poorly that Chevrolet twice suspended production in 2012. "It's clear there's interest in electric vehicles," says Mike VanNieuwkuyk, the executive director of global automotive for the global marketing information services firm J.D. Power and Associates. "But despite the interest, most drivers can't justify the price premium."[4]

The production costs for hydrogen-powered vehicles are even higher. The US Department of Energy (DOE) reports, "Fuel cell vehicles are currently far too expensive for most consumers to afford, and they are only available to a few demonstration fleets."[5] Given the difficulties of finding an onboard hydrogen storage system, this is unlikely to change in the near future.

> "It's clear there's interest in electric vehicles, but despite the interest, most drivers can't justify the price premium."[4]
>
> —Mike Van Nieuwkuyk, executive director of global automotive for J.D. Power and Associates.

Questionable Fuel Savings

One of the ways that manufacturers of alternative energy vehicles have attempted to attract consumers is by touting the energy savings they will accrue over the life of the vehicle. Although hybrids and pure electrics do in fact save owners money at the pump, the initial price differential far exceeds the savings in fuel costs. Statistics indicate that it takes at least eight years for hybrid vehicles to pay back on the investment and fifteen years to pay back on electric vehicles. A study published by the American

Electric-Drive Cars Cost Too Much

Electric cars and hybrids cost more than similar gasoline- or diesel-powered models. This chart shows the 2013 average price paid for various models. Particularly striking are price differences where manufacturers have come out with the same model of vehicle available either with a conventional engine or as pure electric or hybrid. For instance, the Ford Focus, which runs on gasoline, costs significantly less than Ford Focus Electric. More price comparisons are shown for smaller vehicles because there are more plug-in electric and hybrid models in this category.

Small		
Honda Fit	Gasoline	$16,060–$20,384
Ford Focus	Gasoline	$16,683–$24,114
Honda Civic	Gasoline	$17,988–$24,488
Chevrolet Cruze	Diesel	$18,122–$25,448
Honda Insight	Hybrid	$19,177–$24,311
Volkswagen GTI	Gasoline	$23,732–$30,490
Nissan Leaf	Plug-in EV	$27,773–$33,402
Ford Focus Electric	Plug-in EV	$38,262–$38,363
Midsize		
Hyundai Sonata	Gasoline	$21,209–$27,370
Ford Fusion	Gasoline	$21,706–$31,513
Toyota Prius	Hybrid	$23,978–$29,213
Hyundai Sonata Hybrid	Hybrid	$25,127–$29,776
Ford Fusion Hybrid	Hybrid	$26,662–$31,304
Chevrolet Volt	Plug-in Hybrid	$38,508–$38,508
Ford Fusion Energi	Plug-in EV	$28,711–$37,395
Luxury		
Mercedes-Benz E-Class	Gasoline	$46,826–$84,931
Porsche Boxster	Gasoline	$48,902–$59,954
Tesla Model S	Plug-in EV	$59,900–$94,900
Porsche Panamera	Hybrid	$73,695–$166,122
Mercedes-Benz S-Class	Hybrid	$84,744–$197,420

Source: *U.S. News & World Report*, "Best Cars," 2013. www.usnews.rankingsandreviews.com.

Council for an Energy-Efficient Economy found that the Ford Focus electric and Chevy Volt had total lifecycle costs that were $10,000 higher than their conventional counterparts. John Peterson, a lawyer who specializes in the energy market, says, "Unless and until the technology premium falls to a point where the incremental capital investment per gallon of annual gasoline savings is competitive with an HEV, plug-ins will only appeal to a niche market of philosophically committed and mathematically challenged buyers."[6] VanNieuwkuyk says that gas prices would have to increase to six or seven dollars per gallon to make electrically powered vehicles make sense economically.

This is unlikely to happen anytime in the near future. The dire warnings about current oil supplies are exaggerated, and new supplies continue to be found. The CIA estimates Canada's crude oil reserves to be 173.6 billion barrels. This means that Canada has more reserves than any other country in the world except for Venezuela and Saudi Arabia. Some experts suggest that it makes more sense to invest in the Keystone Pipeline to transport oil from Canada to refineries in Texas and other parts of the United States than to invest in finding new ways to power vehicles. Senator John Hoeven of North Dakota says, "The Keystone XL pipeline represents a big step toward true North American energy independence, reducing our reliance on Middle Eastern oil and increasing our access to energy from our own nation and our closest ally, Canada, along with some oil from Mexico—to 75% of our daily consumption."[7]

The Cost of Maintenance and Repairs

Another consideration is maintenance and repair costs. Hybrids have all the routine maintenance costs of conventional internal combustion engine vehicles, with added costs associated with the electrical systems. And while non-hybrid electric vehicles do not have expenses related to the internal combustion engines, they have complex and expensive motor systems and battery packs.

The batteries used in electric cars are not as durable as an internal combustion engine; they will need to be replaced one or more times during the lifetime of the vehicle. If something goes wrong with a battery, the entire battery pack may need to be replaced. Given the fact that the typical battery pack runs from $10,000 to $20,000, this makes electric car ownership an expensive proposition.

Other costs accrue because there are so few electric cars on the road. According to estimates from the EIA's 2013 *Annual Energy Outlook,* just 2.1 million hybrid and electric vehicles were on US roads in 2011, compared to over 123 million conventional cars. This means that the cost of repair and replacement parts will be higher. In addition, fewer certified mechanics will be trained in the repair of alternative energy vehicles, which will drive up the labor costs involved with maintenance and repair. If something goes wrong with the electrical motor, speed controller, or battery systems, the dealership may be the only option for getting the car fixed.

Higher Lifecycle Costs

Hybrid vehicles and electric cars lose value, or depreciate, faster than conventional vehicles. In 2012 the National Automobile Dealers Association (NADA) estimated that plug-in electric cars depreciated by an annual rate of 31.5 percent, compared to just 12.4 percent for conventional automobiles. The rate of depreciation is part of a car's overall value. Electric cars depreciate more quickly in part because of their higher price, but the constant improvements in the technology also play a role. This year's models have better technology than last year's, and next year's model will likely be better yet. Not only that, but the price is coming down considerably, and no one will buy a used car for more than a new one. NADA concludes that "for the foreseeable future, it remains that the value proposition of a plug-in EV will be substantially worse than that of its two counterparts [internal combustion engine and hybrid vehicles]."[8]

> "For the foreseeable future, it remains that the value proposition of a plug-in EV will be substantially worse than that of its two counterparts [internal combustion engine and hybrid vehicles]."[8]
>
> —National Automobile Dealers Association.

In the end, it is the lifecycle costs that matter, and electric cars and hybrids are simply more expensive. According to a 2012 analysis by the Congressional Budget Office, a typical plug-in hybrid electric car's lifetime cost is roughly $12,000 higher than a gasoline-powered car. Electric cars depreciate faster, and the lifecycle of their expensive batteries is uncertain. Given the miniscule number of cars in production, the costs are unlikely to be competitive enough for the average consumer for many decades to come.

Alternative Energy Cars Are Increasingly Affordable

"The public will accept and embrace electric vehicles; some people already have. And the rest of them will come around when technology advances electric vehicles to the point where they offer comparable performance at comparable prices. We'll get there. We will see the day when we have an affordable electric car that offers 300 miles of range with all the comfort and utility of a conventional vehicle."

—Mark Reuss, president of GM North America.

Quoted in *GM News*, "Mark Reuss Remarks to Automotive News World Congress," January 16, 2013. http://media.gm.com.

Given the revolutionary technology used by electric vehicles, it is no surprise that the first models on the market cost more than cars that have been sold for decades. But there is no reason to believe that the costs will remain high. New technologies almost always cost a lot initially, before they are widely used, "and this is no less true for electric cars,"[9] says Elon Musk, the CEO of Tesla Motors. Tesla has promised to put earnings back into research and development to improve upon the technology to ensure that its next model is less expensive.

Electric car manufacturers say that as more cars are made, the costs will come down. With any manufactured item, per-unit costs are much higher when there are fewer goods produced. This is because there are fixed costs—costs that remain the same regardless of how much of something is made. The manufacturing plant is an example of a fixed cost; regardless of how many cars are built, the cost of the plant remains the same. In addition, it is less expensive to buy components in bulk. John Viera, Ford's global director of sustainability, similarly suggests that sales will drive down costs. "There's a lot of component commonality among [electric and hybrid vehicles], so the more hybrids we sell, the cheaper it's going to be to make pure electrics,"[10] he explains.

Production costs for electric and hybrid vehicles have already come down significantly. In large part this is due to improvements to the batteries, which remain the most expensive component. "We have made great strides in reducing costs as we gain experience with electric vehicles and their components,"[11] explains Don Johnson, the US vice president of Chevrolet sales. GM executive Dan Akerson expects manufacturing changes to reduce the cost to build the next-generation Volt, scheduled to be released in 2015, by $7,000 to $10,000.

> "Almost any new technology initially has high unit cost before it can be optimized, and this is no less true for electric cars."[9]
>
> —Elon Musk, CEO of Tesla Motors.

Lower Energy Costs

To determine the relative affordability of alternative energy vehicles, the total cost of ownership must be taken into consideration. The most obvious way that alternative energy vehicles save money is at the pump. The Union of Concerned Scientists (UCS) estimates that the average owner will spend more than $22,000 on fuel over the life of a gasoline-powered car. These calculations assume that the price of gasoline will rise at the same rate it has been rising—an assumption that some say is unrealistic. Given the rising demand for oil and the limited supply, gasoline prices of five or ten dollars a gallon—or more—may be on the horizon. Prices also have become more volatile. "Oil price predictions used to be about oil consumption and markets, but now it's about where the next riot will break out," said Stale Tungesvik, a senior executive at a Norwegian oil company. "It's so much more politically based, and that makes it a mystery to everyone."[12]

Even at today's gas prices, the cost per mile of an alternative energy vehicle is a fraction of that for a gasoline-powered car. Some electric car

Considerable Fuel Cost Savings

The purchase price of electric cars may be higher, but some studies show that they will save money over the long run. Research undertaken by the Union of Concerned Scientists shows that electric vehicles can save an owner almost $13,000 a year over gasoline-powered vehicles.

Lifetime Gasoline Consumption and Fuel/Charging Costs

Type of Vehicle	Gasoline Consumption (in gallons)	Fuel/Charging Costs
Electric Vehicles	0	$5,200
Gas Hybrid Vehicles	3,300	$9,800
Gas-Powered Vehicles	6,100	$18,000

Source: Union of Concerned Scientists, *State of Charge: Electronic Vehicles' Global Warming Emissions and Fuel-Cost Savings Across the United States*, June 2012. www.ucsusa.org.

manufacturers claim that their vehicles require as little as one to two cents a mile to operate, compared to an average minimum cost of about eight to ten cents per mile for even the most fuel-efficient gasoline engine vehicles. The DOE has estimated that owners of electric cars spend only $1.14, on average, to go as far as owners of gasoline-powered cars can go on one gallon of gas. A typical electric vehicle can run for forty-three miles on a dollar's worth of electricity. With gasoline at roughly $4.00 a gallon, even the most fuel-efficient gasoline-powered vehicles can go only about ten miles on a dollar.

Comparing costs of electric and conventional cars requires knowing what one is spending for electricity, which can vary greatly from one place to another. In areas where electricity is inexpensive, energy savings may amount to thousands of dollars over the lifetime of the vehicle. Some electric companies offer lower rates at night, which is usually when owners of plug-in electric cars power their vehicles. Further savings may be achieved if recharging stations allow owners to recharge their batteries for free. Another option may be stations paid for by companies that run their advertisements as drivers recharge their batteries.

Savings on Maintenance and Repairs

Cost comparisons need to also take into account maintenance needs. In contrast to internal combustion engines, which have hundreds of moving parts, electric cars have just a dozen or so parts. The motors are simple, and little can go wrong with them. Electric vehicles do not have the maintenance costs associated with internal combustion engines: They do not require oil changes or tune-ups and do not have spark plugs, radiators, fuel filters, or other equipment that may need to be replaced in internal combustion engines. There is also evidence that brakes get less wear due to the regenerative braking feature. One owner writes of his hybrid, "One of the neat things about a hybrid is that the gas engine is not running when you are stopped or driving slowly. It is amazing how often that happens in city driving. The result is that you are putting less wear on your engine."[13]

The biggest expense is the battery. Critics argue that battery replacement costs outweigh any other cost savings, but the new nickel-metal hydride (NiMH) and lithium-ion (Li-ion) batteries used in today's electric vehicles last much longer than their predecessors. In an evaluation of Toyota's RAV4 EVs conducted by Southern California Edison (SCE), the batteries typically lasted more than one hundred thousand miles in real-world use. The authors of the report on the findings of this evaluation write, "SCE's positive experience points to the very strong likelihood of a 130,000 to 150,000-mile Nickel Metal Hydride battery and drive-train operational life. EVs can therefore match or exceed the lifecycle miles of comparable internal combustion engine vehicles."[14] And when the battery does ultimately run out, the replacement options

> "EVs can . . . match or exceed the lifecycle miles of comparable internal combustion engine vehicles."[14]
>
> —Southern California Edison.

will likely be more inexpensive, lighter, and have more energy capacity than today's batteries.

Costs to Society

Any cost comparison should consider the costs to society as a whole. The costs of alternative energy vehicles come down considerably when taking into account intangible costs of conventional cars, such as those resulting from the long-term impact of emissions. In a 2011 report the Harvard School of Public Health put the health and environmental costs of fossil fuels at $523 billion a year.

Other costs of conventional automobiles relate to the supply chain of oil, including transport, the protection of supply lines, the prevention of oil spills, and costly cleanup when spills occur. British Petroleum (BP) spent more than $25 billion in clean up and restoration costs following the April 2010 oil spill in the Gulf of Mexico. Experts say there have also been incalculable costs to the fishing industry, tourism, and other businesses along the coast.

Switching to electric automobiles will not negate the world's dependence on oil, but it will reduce it. And the fact remains that oil is a limited resource and is therefore subject to significant increases in price. The cost of electricity, meanwhile, is likely to decrease as new renewable energy sources are tapped. The costs of the vehicles themselves and the energy that fuels them will decrease rapidly over the next few years, making alternative energy vehicles increasingly more affordable than their gas-guzzling alternatives.

Chapter Two

Are Alternative Energy Cars a Viable Option?

Alternative Energy Cars Are a Viable Option

As gasoline becomes increasingly expensive, consumers will turn to cars powered by alternative sources. Electric vehicles and fuel-cell technologies are already making their way onto the showroom floor. The world's best engineers and scientists are working on improving batteries to extend the range of electric vehicles and to lessen the time needed for recharging. Electric vehicles will win over consumers with their superior performance, quiet ride, and—eventually—their extended range.

The Debate

Alternative Energy Cars Are Not a Viable Option

Gasoline and diesel are an inexpensive, reliable source of energy that have powered automobiles for more than a century. Engineers have found ways to make them increasingly fuel efficient, and there is already an extensive fueling network in place, ensuring that people can fill up their tanks when and where they need to. None of the technologies for electric vehicles are proven, and the cost remains out of reach. While electric vehicles might be a viable option as a second car for the very wealthy to get around town, they will not displace the internal combustion engine.

Alternative Energy Cars Are a Viable Option

"For electric vehicles (EVs), the future is here. No longer just concept models, EVs are being featured in—and rolling out of—showrooms across the country."

—Union of Concerned Scientists, an independent nonprofit group that works toward improving the health of the environment.

Union of Concerned Scientists, *State of Charge: Electric Vehicles' Global Warming Emissions and Fuel-Cost Savings Across the United States*, June 2012.

In a 2012 survey 62 percent of leading automotive executives said that the use of alternative fuel technologies was "extremely" or "very" important to consumers. And where consumers go, the automakers are sure to follow. Automakers have invested billions in technologies to bring the concepts to fruition. "Thankfully, it takes more than a crusher to kill a technology," writes Chris Paine, harking back to the demise of the electric vehicle in the 1990s, when many ended up being crushed by their manufacturers. "Today, almost all the major automakers, along with a host of new players, are investing in and building plug-in cars."[15]

Concept Cars and Production Vehicles

Alternative energy vehicles may be a rarity on the road, but at car shows across America and the world new concept cars and prototypes are being unveiled. (Automobiles go through many stages before being manufactured, and concept cars and prototypes are used to introduce and/or test new features or technologies.) Indeed, almost all major auto companies have plans for electric models of one kind or another. Carlos Ghosn, the CEO of Nissan and Renault, estimates that battery-powered vehicles will account for 10 percent of global new-car sales by 2020. "The electrification of the vehicle fleet is a foregone conclusion,"[16] announces former GM vice chairman Bob Lutz.

Winning over consumers will be an important part of making alternative energy vehicles, and there is already evidence that this is happening. In just ten years the hybrid gasoline-electric car has gone from curiosity to mainstream. Toyota Prius, the most popular brand, has sold more than a million cars globally and, in 2013, was the third best-selling car in the United States. One of the main reasons for the success of hybrids is that they run the same way as conventional vehicles, requiring no behavioral change on the part of drivers. This may prove to be the path to all-electric vehicles. As engineers find ways to shift more functions to battery power, they may create a gradual, seamless transition from a gasoline-powered vehicle that gets an extra boost from an electric motor to an electric-powered vehicle that relies on gasoline only as an emergency backup power source.

> "The electrification of the vehicle fleet is a foregone conclusion."[16]
>
> —Bob Lutz, former vice chairman at General Motors.

Consumer hesitation is already dissipating. In a Consumer Affairs poll undertaken in the summer of 2012, more than one-third of respondents (37 percent) ranked fuel economy as their top priority for a new car, making this the top-ranking consideration, and 73 percent of drivers said they would consider an alternatively fueled vehicle. The poll showed that younger buyers tend to be more interested in alternative energy vehicles than older buyers are, suggesting that some of the hesitation to embrace new technologies may dissipate in the not-too-distant future.

The Performance Advantage

Electric motors provide several advantages over internal combustion engines. Internal combustion engines produce peak horsepower and torque within a narrow range, requiring an elaborate system of gears (the transmission) to operate. This requires a clutch to propel the vehicle from a stop, a throttle to regulate the speed of the engine, and gears to channel torque to the drive wheels. In contrast, electric motors provide maximum torque at any speed, providing strong acceleration from a stop. The lack of gear-shifting also results in a quieter and smoother ride. The combination offers consumers the quiet ride of a luxury vehicle with the performance of a sports car.

Recent electric models have helped win over skeptics. "Before we were able to get the Roadster out, they'd say you couldn't possibly make the car work," says Elon Musk about his first electric model. "When we did, they said: Well, nobody's going to buy it. And people did."[17]

Easy Charging and Recharging

The main hurdle—range—is not as great as critics assume. The distance that plug-in electric vehicles can travel without needing to be recharged far exceeds the average number of miles the average American travels per day. Easy-to-use home recharging systems allow drivers to conveniently recharge the battery as the car is parked in the driveway or garage. And owners of electric cars will never again have to go to a filling station.

For longer-distance travelers, electric-car proponents envision a series of recharging stations. John Voelcker, who produces *Green Car Reports*, writes that electric vehicles "have at least the basic of a widespread 're-fueling' system in place already: the electric grid."[18] The recharging stations that are popping up in places around the world allow cars to be charged much faster than at home and sometimes completely free of charge. There are already more than fifty-seven hundred public stations in the United States, many with several charging points. Tesla Motors has worked with a major hotel chain to have recharging stations available for patrons to recharge their vehicles overnight and has promised to set up a network of fast-charging stations nationwide. Following the introduction of the battery-powered Leaf in the fall of 2010, Nissan similarly announced plans to work with private companies to set up charging station networks where its vehicles are sold.

For those who do not want to wait twenty to thirty minutes for their battery to be charged, electric car proponents suggest an alternative to

Consumers Want Electric Vehicles

As shown in this chart, 36 percent of respondents in a 2012 survey said they are "extremely" or "very" interested in purchasing a plug-in electric vehicle. Other studies suggest that younger people tend to be more open to these alternative energy options. As production brings the price down and people become more confident with the reliability and safety of electric cars, these consumers will make the electric vehicle the car of the future.

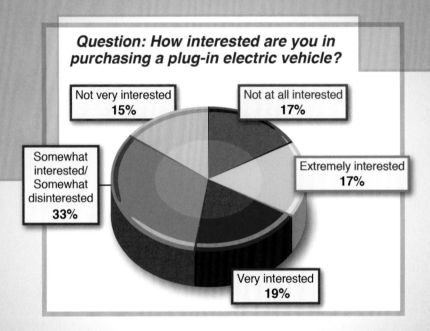

Question: How interested are you in purchasing a plug-in electric vehicle?

Not very interested
15%

Not at all interested
17%

Somewhat interested/ Somewhat disinterested
33%

Extremely interested
17%

Very interested
19%

Source: Green Car Congress, "Pike Research US Consumer Surveys Finds Decreasing Fundamental Interest in Plug-In Electric Vehicles," October 29, 2012. www.greencarcongress.com.

recharging stations: stations where drivers could swap their drained battery for a fully charged alternative. In the summer of 2013 Musk demonstrated the viability of this approach, swapping the battery in roughly thirty seconds—much faster than it would take to fill up a tank with gasoline.

Better Batteries

Automakers are also seeking to address the problem of range by finding newer, better, more powerful batteries. "The hurdle between the well matured internal combustion engine technology and hybrid electric vehicle and electric vehicles to take its rightful place as a successor is a high-energy, high-density storage system," writes Etim U. Ubong, an associate professor and director of the PEM Fuel Cell Program at Kettering University. "Within the next decade, the research community and the manufacturers of optimized electrical devices will perfect this technology and ICE [internal combustion engines] will gradually give way to hybrids."[19]

Researchers also are making great strides in hydrogen fuel cell technologies, which then US president George W. Bush touted in 2003 as "one of the most encouraging, innovative technologies of our era."[20] Fuel cell technologies have been used to power robotics, including the Lunar Rover that explored the moon decades ago. Authors of the Green Fuel Online website say the technology is far more accessible than it may appear: "As farfetched as it might seem, [running cars with hydrogen] is very much a possibility if one takes into account that vehicles such as airplanes, submarines and space shuttles already run by hydrogen at least partially. . . . Spacecrafts, weather stations, rural locations and military submarines are a few examples of where fuel cells are currently employed for electricity generation."[21]

Mercedes-Benz, Ford, and Nissan are all working on fuel cell technology, and Toyota has announced plans to launch a hydrogen fuel cell sedan in 2015. "Compared to their early application in the Gemini space missions in the 1960s, fuel cells have become astonishingly small and powerful, evolving rapidly since vehicular applications began in earnest,"[22] says Toyota.

> "Compared to their early application in the Gemini space missions in the 1960s, fuel cells have become astonishingly small and powerful, evolving rapidly since vehicular applications began in earnest."[22]
>
> —Toyota, *Toyota FCHB* (Fuel Cell Hybrid Vehicles) *Book.*

Whether the future will belong to the plug-in electric, fuel-cell hydrogen, or hybrid vehicles remains to be seen. Given the millions of dollars governments, automakers, and other businesses are investing in alternative technologies, the deciding factor will not be what is possible but rather what is most desirable.

Alternative Energy Cars Are Not a Viable Option

"Because of its shortcomings—driving range, cost and recharging time—the electric vehicle is not a viable replacement for most conventional cars."

—Takeshi Uchiyamada, vice chairman of Toyota and "father" of the Prius.

Quoted in Norihiko Shirouzu, Yoko Kubota, and Paul Lienert, "Are Electric Cars Running Out of Juice Again?," *Reuters*, February 4, 2013.

Range is the biggest issue plaguing engineers of electric cars and the biggest obstacle to their adoption by the American public. In a 2011 Gallup poll, 57 percent of respondents said they would not buy an electric car regardless of the price of gas. "Our research shows that people want to feel like they can get into their car and drive across the country . . . if they have to," says an executive at Better Place, an international company founded to sell battery-charging services. "It might sound silly, but it's real."[23]

Inherent Limitations of Batteries

For decades, engineers have sought a battery system that could overcome issues of range, but they have yet to succeed. Regardless of their size, batteries can only store a finite amount of energy; when the energy is exhausted, they need to be recharged. And while the typical automobile can go roughly three hundred miles before needing fuel, engineers have not found a battery that can come even close to this range—at least without making the size and weight unfeasible for an automobile. "We've always got the discussion between the weight of the battery, the cost of the battery, and how far people want to go," explains Tom McCabe, a senior engineer at Nissan. "For example if we gave you a car with 500 miles, it would be a very heavy car, it would be a very expensive car and we'd also have to completely change the package."[24]

> ## "When a battery car's range gets in the Leaf zone . . . you can't even give it away."[27]
>
> —Patrick Michaels, director of the Center for the Study of Science at the Cato Institute.

The problem is not just an electric car's range, it is also the fact that the range can be highly variable, depending on everything from the temperature outside, to the number of people inside, to driving conditions and characteristics. The resulting uncertainty, coupled with the long recharging time, creates what columnist Louis Woodhill refers to as "range anxiety." Even those with short daily commutes may justifiably worry about getting stuck in a snowstorm or traffic jam. "Reviews of the Leaf are filled with accounts of drivers turning off the A/C in the summer and the heat in the winter," reports *Forbes* magazine. "Some drivers even decided that they couldn't risk charging their cell phones, using the radio, or turning on the windshield wipers."[25]

Recharging is inconvenient at best. While filling up a tank with conventional fuel takes a few minutes, fully recharging an electric car's battery pack takes hours. Even with the best, most expensive battery pack and auxiliary charging system on the market today, a "quick charge" to 50 percent capacity takes at least thirty minutes. And while manufacturers and proponents of electric vehicles have pointed out that electric charging stations could replace today's gas stations, the infrastructure could take decades to get in place. According to the National Renewable Energy Laboratory, eleven states had no public charging stations at all as of the summer of 2011, and sixteen states had ten or fewer.

Sluggish Sales

There can be no move to alternative energy vehicles if consumers are unwilling to buy them, and that has proved to be the case. Sales of the Chevrolet Volt, a plug-in hybrid introduced by General Motors in 2012, have been consistently under projections; the company suspended production twice in 2012 because of the lack of demand. "The reality is that consumers continue to show little interest in electric vehicles,"[26] conclude the authors of a 2013 article published by Reuters.

Under pressure to meet California's quotas for zero-emission vehicle sales, some automakers have offered lease deals on electric cars for less than $200 a month, but even at rock-bottom prices, there are few takers. Chevrolet sold fewer Volts in the entire month of June than it did large pick-up trucks in a single day. Nissan says that its Leaf can go seventy-five miles on a single battery charge, but many people say this is a stretch. Regardless, it appears to be too few miles for most consumers. Patrick Michaels, the director of the Center for the Study of Science at the Cato Institute, writes, "When a battery car's range gets in the Leaf zone . . . you can't even give it away. The federal [tax rebate of] $7500 with [an additional] $2500 in several states, and the reduced fuel costs, more than pay for two-year leases on the car. That's right, free transportation, and sales still [are dismal]."[27]

Manufacturers have lowered their price on electric vehicles, but many experts believe that consumers will not want electric vehicles at any price. "There are tremendous obstacles from an economic outlook, even with the price decreases," concedes John O'Dell, an analyst for the automotive information source Edmunds. O'Dell specializes in environmentally friendly cars

> "The price decrease isn't going to open up the floodgates for EV sales. Honestly, I'm not sure if the floodgates ever do open for battery electrics."[28]
>
> —John O'Dell, an analyst who specializes in environmentally friendly cars.

and personally owns—and loves—his first-generation Nissan Leaf. "The price decrease isn't going to open up the floodgates for EV sales. Honestly, I'm not sure if the floodgates ever do open for battery electrics."[28]

Hydrogen Fuel Cells

Hydrogen power, long assumed to be the fuel of the future, has many of the same problems as electricity and then some. Fuel cell systems have

Electric Vehicles Cannot Meet Consumer Needs

Range is a major problem for electric cars. Even the best of them with the most expensive batteries cannot go much farther than one hundred fifty miles before needing to be recharged. The dearth of public electric vehicle charging stations makes the car impractical. As shown here, only six states have more than fifty public stations, and ten states do not have any.

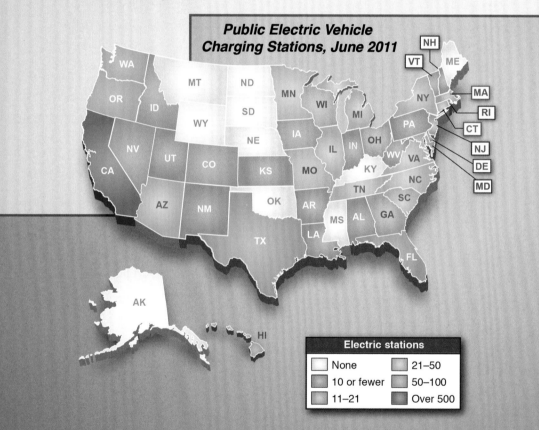

Public Electric Vehicle Charging Stations, June 2011

Electric stations

- None
- 10 or fewer
- 11–21
- 21–50
- 50–100
- Over 500

Source: US Department of Energy, "Careers in Electric Vehicles," Bureau of Labor Statistics, June 2011. www.bls.gov.

proven to be too fragile to endure the bumps and vibrations to which motor vehicles are subject. Engineers also have not found ways to reliably overcome problems related to freezing. Daimler-Benz announced that it was investing in fuel-cell development almost two decades ago, promising to sell forty thousand fuel cell cars by 2004, but its plans have not come to fruition. More recently, Mercedes teamed up with Ford and Nissan on fuel-cell development, investing upward of $100 million per year, but plans to offer a hydrogen-powered production car are years behind schedule. Some analysts say that may never happen.

One of the problems is that automakers cannot recoup their costs of production without offering the cars at a price that the market is unwilling to pay. Even if problems related to the fuel cells can be resolved, the infrastructure for providing hydrogen fuel will be far too costly. Hydrogen is difficult to transport and store and would require an entirely new distribution system. It costs almost $2 million to build a hydrogen fueling station, and there is no incentive for oil companies or other possible providers to take on the mammoth task of building a network of stations. Analysts say that at least fifteen thousand such fueling stations (if they were ideally spaced) would be needed to cover the United States. Hydrogen fueling stations do not fall under any sections of existing municipal zoning codes, yet another obstacle for the development of a so-called hydrogen highway.

> "I do not currently see a situation where we can offer fuel cell vehicles at a reasonable cost that consumers would also be willing to pay."[29]
>
> —Martin Winterkorn, CEO of Volkswagen.

Martin Winterkorn, Volkswagen's CEO, is among those who believe that these obstacles will prove to be too significant for hydrogen fuel cell vehicles to become a reality. "I do not see the infrastructure for fuel cell vehicles, and I do not see how hydrogen can be produced on large scale at reasonable cost," Winterkorn said at a 2013 press conference. "I do not currently see a situation where we can offer fuel cell vehicles at a reasonable cost that consumers would also be willing to pay."[29]

Looking into a Crystal Ball

The Committee on Transitions to Alternative Vehicles and Fuels admits that it cannot predict which technologies will take off. "Many technologies, with widely varying levels of current capability, cost, and commercialization, can reduce light-duty vehicle petroleum consumption, and most of these also reduce greenhouse gas emissions," writes committee chair Douglas M. Chapin. "However, any transition to achieve high levels of reduction is likely to take decades."[30]

Chapter Three

Do Alternative Energy Cars Benefit the Environment?

Alternative Energy Cars Are Good for the Environment

Throughout the world transportation represents a significant source of carbon dioxide, a greenhouse gas that contributes to global warming. Automobiles also contribute to air pollution and smog, particularly in parts of the world with few air quality regulations and/or inadequate oversight. A shift to alternative energy vehicles is an essential step toward minimizing the harmful impact of transportation on the environment.

The Debate

Alternative Energy Cars Are Not the Answer to Environmental Concerns

Alternative energy vehicles are not the environmental panacea that proponents say they are. Interest in electrical vehicles stems from the fact that they offer no tailpipe emissions, but if the electricity is generated from fossil fuels, they may be just as polluting—or even more so. If the grid is powered by coal, for instance, the net impact may be harm to the environment. Production of the vehicles and the huge batteries needed to power them adds to their negative environmental impact.

Alternative Energy Cars Are Good for the Environment

"From cradle to grave, electric cars are the cleanest vehicles on the road today."

—Max Baumhefner, Sustainable Energy Fellow, National Resources Defense Council.

Max Baumhefner, "Electric Cars Are Cleaner Today and Will Only Get Cleaner Tomorrow," The Energy Collective, August 12, 2013. http://theenergycollective.com.

Decades ago, the clouds of smog that enveloped cities like Los Angeles propelled policy makers to enact stricter emissions standards. US engineers responded with considerable improvements to the internal combustion engine. Today's cars emit almost no carbon monoxide, hydrocarbons, nitrogen oxides, or other pollutants that have been blamed for smog.

Unfortunately, engineers have not been successful in reducing carbon dioxide emissions. This is because it is an unavoidable by-product of the internal combustion engine, in which fossil fuels (usually gasoline or diesel) are used to create a series of explosions. It is this series of explosions that creates the energy to power the vehicle. The explosions also create carbon dioxide, a waste by-product that is pumped out through the car's exhaust system. For every gallon of gasoline burned in the average car, roughly twenty pounds of carbon dioxide is released into the air. EPA's "well-to-wheel" calculations, which factor in the costs of acquiring, processing, transporting, and distribute fuel, indicate that the average American vehicle is responsible for six metric tons of carbon dioxide emissions—approximately three times the weight of the average car.

The negative impact of carbon dioxide is perhaps less noticeable than smog, but carbon dioxide may have an even greater impact on the environment over the long term. Carbon dioxide is a greenhouse gas that gets stuck in the atmosphere, making it harder for the sun's radiation to escape and causing temperatures on Earth to increase. Scientists warn that the resulting global warming could cause a wide variety of problems, including drought, extreme weather, and rising sea levels.

Alternative energy vehicles are an important means of reducing our carbon footprint, defined as the sum of greenhouse gas emissions from various activities. Studies show that hybrids can reduce carbon dioxide emissions by 50 percent and carbon monoxide by 90 percent. Transitioning to plug-in electric vehicles offers even greater environmental benefits. In a 2012 study published in the *Journal of Industrial Ecology*, the authors conclude that "the combination of EVs with clean energy sources would potentially allow for drastic reductions of many transportation environmental impacts, especially in terms of climate change, air quality, and preservation of fossil fuels."[31]

> "The combination of EVs with clean energy sources would potentially allow for drastic reductions of many transportation environmental impacts, especially in terms of climate change, air quality, and preservation of fossil fuels."[31]
>
> —Troy R. Hawkins, Bhawna Singh, Guillaume Majaeu-Bettez, and Anders Hammer Strømman, researchers at the Norwegian University of Science and coauthors of a comparative study on electric vehicles.

Electricity and the Environment

For plug-in hybrids and electric vehicles, the environmental impact depends on the power source. But for most, the electric vehicle comes out ahead. According to a 2013 study, 83 percent of Americans live in areas in which charging an electric vehicle on the electricity grid emits global warming pollution that is less than or equal to the best hybrids on the road. Where electricity is generated from the sun, wind, or water, electric cars have no impact on the environment. And the Electric Vehicle Association of Canada (EVAC) says that even when electricity is generated by coal, electric vehicles cut carbon emissions roughly in half. In its 2012 report the UCS concludes that "consumers should feel confident that driving an electric vehicle yields lower global warming emissions than the average new compact gasoline-powered vehicle."[32]

Moreover, the environmental advantages of electric vehicles over conventional cars will widen as electric companies increasingly look to wind, water, and other renewable energy sources for power. "The longer you own an electric vehicle, the *lower* its global warming emissions are likely to become," writes Don Anair, deputy director of the Clean Vehicles program at the Union of Concerned Scientists. "As some of the oldest, dirtiest coal plants are being retired and investments in renewable electricity increase, the global warming emissions that result from generating a given amount of electricity are estimated to fall nationwide by an average of about 13 percent by 2025."[33] The US Energy Information Administration (EIA) projects that renewable energy sources will grow from about 11 percent of the total energy production in the United States in 2009 to about 15 percent in 2025.

> "Consumers should feel confident that driving an electric vehicle yields lower global warming emissions than the average new compact gasoline-powered vehicle."[32]
>
> —Union of Concerned Scientists.

Individual owners can improve the potential environmental benefits of electric vehicles. Some proponents of electric vehicles recommend adding solar panels to generate the electricity to fuel them. This would negate any environmental impact of owning the car. Consumers who have a choice of power providers can maximize the green factor by choosing a provider that supplies electricity from renewable sources or that has a green power program that sets aside funding to support cleaner energy.

Reducing Dependence on Nonrenewable Energy

Electric vehicles not only keep the environment cleaner, they also help to reduce the United States' dependence on oil. The UCS warns, "When it comes to oil use, our country is at a crossroads: we can put the United States on a path toward cutting projected oil consumption in half, or we can continue to threaten our health and economic well-being by moving to increasingly dirty and inaccessible (and therefore

Future Alternative Energy Vehicles Will Reduce Greenhouse Gases

This chart shows projected amounts of greenhouse gases that will be generated by vehicle types in 2035–2045. The data is drawn from well-to-wheels analyses that consider all steps of the energy chain, from fuel extraction/production to vehicle use. The first bar shows that today's vehicles spew roughly 450 grams of carbon dioxide per mile, which is much higher than alternative energy cars of the future. Even when including the greenhouse gases emitted to produce electricity, conventional internal combustion vehicles are responsible for almost twice the greenhouse gas emissions as a battery electric vehicle and as much as ten times as much as a fuel cell electric vehicle.

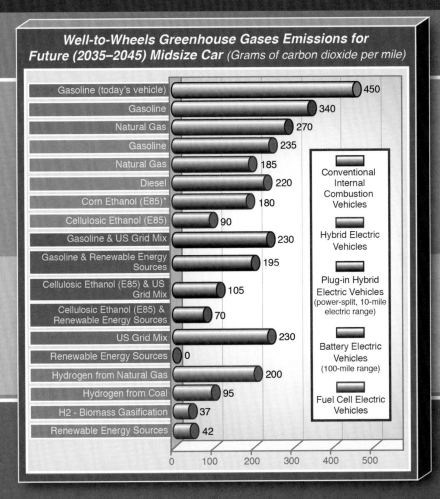

Well-to-Wheels Greenhouse Gases Emissions for Future (2035–2045) Midsize Car *(Grams of carbon dioxide per mile)*

Gasoline (today's vehicle)	450
Gasoline	340
Natural Gas	270
Gasoline	235
Natural Gas	185
Diesel	220
Corn Ethanol (E85)*	180
Cellulosic Ethanol (E85)	90
Gasoline & US Grid Mix	230
Gasoline & Renewable Energy Sources	195
Cellulosic Ethanol (E85) & US Grid Mix	105
Cellulosic Ethanol (E85) & Renewable Energy Sources	70
US Grid Mix	230
Renewable Energy Sources	0
Hydrogen from Natural Gas	200
Hydrogen from Coal	95
H2 - Biomass Gasification	37
Renewable Energy Sources	42

Legend:
- Conventional Internal Combustion Vehicles
- Hybrid Electric Vehicles
- Plug-in Hybrid Electric Vehicles (power-split, 10-mile electric range)
- Battery Electric Vehicles (100-mile range)
- Fuel Cell Electric Vehicles

* E85 is a fuel blend that uses 85 percent ethanol and 15 percent gasoline. Ethanol is derived from plants, and corn is the most common source.

Source: US Department of Energy, "Well-to-Wheels Greenhouse Gas Emissions and Petroleum Use for Midsize, Light Use Vehicles," 2010. www.hydrogen.energy.gov.

dangerous) sources of oil."[34] The organization calculates that the use of electric vehicles could cut US oil use by nearly 1.5 million barrels a day by 2035.

Using less oil has a number of benefits, environmental and political. Among them is the reduced risk of oil spills and their devastating effects. Oil spills can be the result of a problem at any stage of the supply chain, from offshore platforms and drilling rigs, to the release of crude oil carried by tankers, to spills of gasoline and diesel at filling stations. Even minor oil spills wreak havoc on the environment. Spilled crude oil penetrates the plumage of birds, making it impossible for them to fly, and adheres to the fur of marine creatures and mammals, making it difficult for them to swim, stay warm, or even breathe. Even if animals survive the immediate effects, they may suffer the effects of toxic compounds ingested as they clean their own feathers or fur or by eating prey that has been affected. The long-term damage to habitats and breeding or nesting grounds can have a profound impact on species long after the spill has been cleaned up.

> "The longer you own an electric vehicle, the *lower* its global warming emissions are likely to become."[33]
>
> —Don Anair, deputy director of the Union of Concerned Scientists' Clean Vehicles program.

In a report written three years after the BP oil spill in the Gulf of Mexico, the National Wildlife Federation reported that the spill continued to have serious ecosystem-wide effects. "Oil spill disasters have taken years to reveal their full effects, and often recovery is still not complete decades later," write the report's authors. "Nearly a quarter-century after the *Exxon Valdez* spill in Prince William Sound, clams, mussels, sea otters and killer whales are still considered 'recovering,' and the Pacific herring population, commercially harvested before the spill, is showing few signs of recovery."[35]

Materials Usage

A final criticism of electric cars comes from concerns about the materials used in the batteries. Granted, batteries use lithium and rare earth elements, but only in very small amounts. "It is correct to raise awareness that high-tech fixes like . . . electric vehicles use more toxic compounds and processes in their manufacture than is generally understood, but we've known this problem since computer technology skyrocketed fifteen years ago as they have exactly the same problems,"[36] writes James Conca, an environmental scientist. And, whereas lead-acid batteries are still commonly used in conventional vehicles, the advanced batteries in electric vehicles use lithium, nickel, and other elements and usually include neither heavy metals nor toxic materials. Environmental problems stemming from the manufacturing process pale in comparison to the degradation caused by gasoline-powered vehicles over the course of many years.

Unlike lead batteries, almost all of the batteries used in today's electric vehicles are recyclable. After they are retired from life on the road, electric-car batteries can continue to be used as energy-storage devices in less-demanding applications for utilities, businesses, and homes. If the batteries themselves cannot be reused, the cobalt, aluminum, nickel, and other metals inside them can be.

In an era of global warming, the electric vehicle glimmers as a beacon of hope. Studies have shown that, even when powered by electricity generated by nonrenewable resources, the electric car can provide considerable environmental benefits. The ongoing transition to greener sources of energy for electricity will make them increasingly attractive options in the years to come.

Alternative Energy Cars Are Not the Answer to Environmental Concerns

"In the analysis including potential effects related to acid rain, airborne particulate matter, smog, human toxicity, ecosystem toxicity and depletion of fossil fuel and mineral resource, electric vehicles consistently perform worse or on par with modern internal engine vehicles."

—Anders Hammer Strømman, professor at the Norwegian University of Science who specializes in research on environmental assessment of energy and transport systems.

Quoted in BBC News, "Electric Cars 'Pose Environmental Threat,'" October 4, 2012.

Alternative energy vehicles are not the environmental panacea that proponents say they are. Interest in electrical vehicles stems from the fact that they offer no tailpipe emissions, but this is just one of many aspects of an automobile's environmental impacts. John DeCicco, a research professor at the University of Michigan School of Natural Resources and Environment explains, "The missing link for really cleaning up cars is not about the car at all. It's about limiting net carbon impacts in the energy and natural resource sectors that supply motor fuel, whatever form that fuel may take."[37] Some of the ways that fuel is obtained—from being extracted out of the ground and made into electricity—may be more harmful to the environment than others. DeCicco also points out that fuel efficiency in gasoline-powered vehicles is improving by nearly 4 percent a year, compared to less than 1 percent in improvement of emissions in electric power generation. Hence, electric power companies continue to rely on fossil-fuel burning and carbon-emitting sources of energy, while internal combustion engines get cleaner.

Electric Power

Like any car, electric vehicles have to get their power from somewhere. For the most part, the electricity that runs electric cars comes from the same source as the electricity that runs refrigerators, lights, and computers in homes and businesses—power plants. Over 70 percent of US power plants get their energy from fossil fuels such as coal or natural gas. According to a 2011 report by the US Environmental Protection Agency (EPA), electricity generation accounted for 41 percent of the carbon dioxide emissions from fossil fuel combustion in 2009. The environmental report card is even worse in China, where most electricity is produced with coal. "An electric car powered with that electricity will emit 21 percent *more* CO_2 than a gasoline-powered car,"[38] reports Bjørn Lomborg, a Danish environmentalist and author of *The Skeptical Environmentalist*. Power plants, especially coal power plants, may also spew other toxic chemicals into the air. The results are harmful to people and the environment. The American Lung Association attributes approximately thirteen thousand deaths annually to air pollution from America's power plants. In parts of the world with minimal air pollution standards, the hazards may be even worse. Additional environmental degradation results from getting the fossil fuel out of the ground and transporting it to the power plant. These environmental issues are ignored when people focus on emissions.

> "The production of lithium batteries needed to operate an electric vehicle is a far cry from environmentally friendly."[40]
>
> —Brigham McCown, the principal and managing director of United Transportation Advisors.

The batteries on electric vehicles are huge, and recharging uses a significant amount of electricity. Some experts warn that the current electric grid would be unable to support the conversion to electric vehicles. If too many people tried to charge their electric car batteries at the same time, it could overwhelm the electric system, resulting in black-outs or brown-outs. In the end, more power plants may be needed to generate the requisite energy, increasing the use of fossil fuels to power the grid and, consequently, the emission of carbon dioxide and other pollutants.

Electric Cars Are Not Environmentally Friendly

Most people assume that electric cars are environmentally friendly but the facts do not support this assumption. In a 2010 report, the National Academies Press compared the costs and benefits that are associated with the production, distribution, and use of the various types of energy used to power cars. The study looked at the anticipated environmental costs of various fueling options for 2030. As shown in the chart, hydrogen and electricity have a greater environmental impact than gasoline or natural gas.

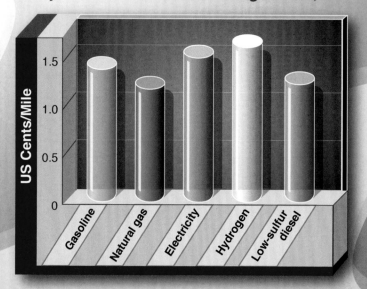

Projected Environmental Damage Costs, 2030

Source: Ozzie Zehner, "Unclean at Any Speed: Electric Cars Don't Solve the Automobile's Environmental Problems," June 30, 2013. www.spectrum.ieee.org.

Proponents of electric vehicles point out that they can be powered by green energy sources, but this solution is neither viable nor environmentally friendly. In a 2013 article, "Unclean at Any Speed," ex-GM employee Ozzie Zehner writes about the downsides of switching to solar energy. The solar cells used for solar energy "contain heavy metals, and their manufacturing releases greenhouse gases such as sulfur hexafluoride, which has 23,000 times as much global warming potential as CO_2, according to the Intergovernmental Panel on Climate Change. What's more, fossil fuels are burned in the extraction of the raw materials needed to make solar cells and wind turbines—and for their fabrication, assembly, and maintenance."[39] Other seemingly green power sources—wind, hydroelectric, and nuclear—have similarly undesirable environmental consequences.

Batteries and Motors

Another environmental concern relates to the materials used in the batteries of electric and hybrid cars. Authors of a 2012 report published in the *Journal of Industrial Ecology* concluded that the production of electric cars resulted in significantly more environmental degradation than conventional vehicles. "The production of lithium batteries needed to operate an electric vehicle is a far cry from environmentally friendly," Brigham McCown, the principal and managing director of United Transportation Advisors, explains in a June 2013 article. "By the time an electric-car leaves the line, it has already produced 30,000 pounds of carbon dioxide emission. A conventional car by comparison only produces 14,000 pounds."[40]

The production of the high-powered batteries used in alternative energy vehicles requires toxic minerals such as nickel, copper, and aluminum. Batteries also use rare earth metals that are recovered through an intensive mining process. The vast majority of rare earth elements are found in China, which lacks environmental regulations or oversight; critics say that the mines leach toxic substances and radioactive waste into nearby land and waterways, posing a threat to the health of humans and the environment. "As the world's hunger for these elements increases, the waste is going to increase," says Nicholas Leadbeater, a chemist at the University of Connecticut who

> "This planet will not be rescued by superexpensive technology for the few, but when the majority of mobility is clean."[42]
>
> —Rainer Michel, the vice president for product planning at Volkswagen of America.

focuses on developing green technologies. "The more mines there are, the more trouble there's going to be."[41]

It also takes considerable energy to mine, smelt, and process the iron, lithium, nickel, and rare earth metals used in electric vehicles, not to mention transporting them from where they are mined in China to processing plants, to factories where they are made into batteries, and to the automotive plants for assembly into the vehicles. By the time they make it into a vehicle on the road, some materials have traveled around the world several times.

There are also end-of-life problems. Some of these batteries are expected to last only five to ten years—far less than the lifetime of the vehicle. Disposing of these huge batteries could create a problem for landfills.

Improving Fuel Efficiency

Some experts say it makes far more sense to focus attention on making improvements in traditional fuels and/or automobile engines than to search for something entirely new. "This planet will not be rescued by superexpensive technology for the few, but when the majority of mobility is clean,"[42] says Rainer Michel, the vice president for product planning at Volkswagen of America.

Significant improvements in fuel economy have been made through efficient transmissions with better torque converters, more gears, the incorporation of electronic engine controls, and computer-controlled fuel injection. Automakers have further improved fuel economy by making cars with sleeker lines that reduce the amount of wind resistance. The result has been a 61 percent improvement in average miles per gallon fuel economy over 1975 levels.

> "The missing link for really cleaning up cars is not about the car at all. It's about limiting net carbon impacts in the energy and natural resource sectors that supply motor fuel, whatever form that fuel may take."[37]
>
> —John DeCicco, a research professor at the University of Michigan School of Natural Resources and Environment.

Better averages could be obtained by getting Americans to give up their trucks and SUVs in favor of smaller, lighter automobiles—a far easier sell than expensive and unreliable alternative energy vehicles. Combined with the integration of flex-fuel options such as ethanol, continued improvements in the design of the automobiles that are on the roads today will continue to reduce their environmental impacts.

Before assuming that alternative energy vehicles are the answer to environmental problems, policy makers and automakers need to make sure that they have fully assessed their entire environmental impact. "To ensure that the promotion of EVs to reduce greenhouse gas (GHG) emissions from transport does not lead to other undesired consequences, it is critical to conduct rigorous, scenario-based environmental assessments of proposed technologies before their widespread adoption,"[43] advise experts at the Norwegian University of Science. Only by looking at their entire environmental impact can policy makers and consumers know the best road to take to a greener future.

Chapter Four

Should the Government Support the Development of Alternative Energy Cars?

The Government Should Support the Development of Alternative Energy Cars

The development of the technologies needed to power vehicles without gasoline or diesel is extremely expensive, and government needs to support these efforts. Government also needs to be involved in educating consumers about the benefits of alternatively powered vehicles and providing incentives for them to change their behavior. It is in society's best interest to reduce the nation's—and the world's—dependence on oil and to provide clean-burning fuels. By helping the manufacturers of automobiles and their components, government is paving the way for a more secure and healthier environment.

The Debate

The Government Should Not Be Involved in Developing Alternative Energy Cars

Government involvement in market decisions often does more harm than good. Experience suggests that government incentives often provide funding for private-sector entities that are not viable without government assistance. As a result, governments have spent millions of dollars funding businesses that went bankrupt and enterprises that failed to deliver on their promises. The success of the Toyota Prius, which came to market with no government assistance, shows that innovation pays off for automakers when they develop an affordable, quality product that consumers want.

The Government Should Support the Development of Alternative Energy Cars

"We need smart government policies that provide incentives for automakers and consumers to invest in clean car technology—and help move America toward a cleaner, safer transportation future."

—Union of Concerned Scientists, an independent nonprofit group that works toward improving the health of the environment.

Union of Concerned Scientists, "Clean Vehicles: Advanced Vehicle Technologies," 2013. www.ucsusa.org.

Governments play an important role in environmental protection, resource conservation, and the ongoing quest for renewable energy. Government actions have also spurred the private sector to find better alternatives to conventional gas guzzlers. In 1990, for instance, when CARB issued a mandate for 2 percent of vehicles sold in the state of California to emit zero pollutants, automakers responded with a number of electric cars.

Lukewarm consumer response to the new electric cars prompted CARB to back off of its mandates. Within months, nearly all the production electric cars were withdrawn from the market. By 2003—the year in which CARB envisioned one in every ten cars to have zero emissions—the generation of electric vehicles inspired by this vision was a thing of the past.

Spurring New Ideas and the Economy

One reason government mandates are needed to spur development of alternative energy vehicles is that research and development costs are astronomical. Without government incentives, the costs would preclude any automaker from making these investments. In a 2013 survey, 79 percent of the world's top automobile manufacturing executives said that government subsidies are needed to enable production of electric vehicles.

Experience further suggests that government investment can be the catalyst for game-changing technologies. In 1993 the DOE began its HEV program as a cost-sharing partnership with General Motors, Ford, and DaimlerChrysler in which the three automakers committed to producing market-ready HEVs within ten years. Threatened by the competition, Toyota also began to explore hybrid technology, coming to market with the Prius in 2000. Similarly, Tesla took advantage of a $465 million federal loan and other financial incentives to introduce its all-electric Model S. In the spring of 2013—nine years before the loans were due—Tesla paid off the government's loan in full. "Tesla's success . . . demonstrates the essential role of private-public partnerships in fostering innovation and bringing new products to market," says Phyllis Cuttino, the director of the Pew Clean Energy Program. She adds, "The cost of an advanced battery has fallen more than 50 percent since 2008 because of the Department of Energy's investment in this technology. Similarly, pilot programs supported by local and state governments have increased the number of charging stations to more than 17,000 across the United States."[44]

In addition to encouraging green transportation alternatives, grants can help support the economy. As part of the American Recovery and Reinvestment Act of 2009, over $2 billion was awarded to firms willing to share the costs for developing and deploying advanced batteries, electric drive vehicles, and other alternative energy options. "By supporting key technologies and sound business plans, we can jumpstart the production of fuel-efficient vehicles in America. We have an historic opportunity to help ensure that the next generation of fuel-efficient cars and trucks are made in America," President Barack Obama said in a 2009 statement. "These investments will come back to our country many times over—by creat-

> "Tesla's success . . . demonstrates the essential role of private-public partnerships in fostering innovation and bringing new products to market."[44]
>
> —Phyllis Cuttino, the director of the Pew Clean Energy Program.

Support for spending government money on development of alternative fuel cars has decreased since the mid 2000s, but nevertheless remains high among adults polled nationally and among people identifying themselves as Democrats. Support during this period has fallen among people identifying themselves as Republicans.

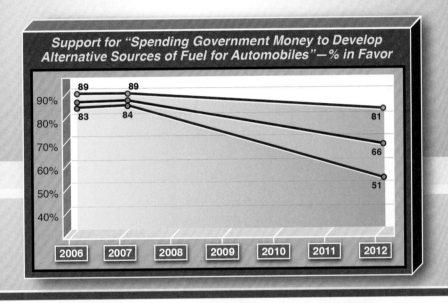

Support for "Spending Government Money to Develop Alternative Sources of Fuel for Automobiles" — % in Favor

Democrats/Dem. leaners   Republicans/Rep. leaners   total in favor

Source: Gallup Politics, "Americans Endorse Various Energy, Environment Proposals," April 9, 2012. www.gallup.com.

ing new jobs, reducing our dependence on oil, and reducing our greenhouse gas emissions."[45]

Investment in alternative energy technologies also stimulates competition among manufacturers of conventional automobiles and components. Executives in the automobile industry recognize that people are looking for fuel efficiency. The increasing threat of competition from hybrids and other alternative energy vehicles has helped fuel the search for better fuel efficiency in conventional cars. As a result, new technologies have resulted in cleaner and more efficient gasoline and diesel engines, as well as the use of ethanol, biodiesel, and other fuels. State and federal policies can also encourage the manufacturing of electric vehicles, batteries, and fuel cells, and stimulate interest in the private sector to ramp up the requisite charging and fueling infrastructure.

Subsidies for Consumers

Government funding can also help encourage consumers to try a new technology. After the introduction of any new technology, consumers tend to wait to see whether it lives up to its promises, particularly with a major purchase like an automobile. "Widespread consumer acceptance of alternative vehicles and fuels faces significant barriers, including the high initial purchase cost of the vehicles and the perception that such vehicles offer less utility and convenience than conventional [internal combustion engine vehicles]," concludes a 2013 National Research Council report. "Overcoming these barriers is likely to require significant government policy intervention that could include subsidies and vigorous public information programs aimed at improving consumers' familiarity with and understanding of the new fuels and powertrains."[46]

> "By supporting key technologies and sound business plans, we can jumpstart the production of fuel-efficient vehicles in America. . . . These investments will come back to our country many times over—by creating new jobs, reducing our dependence on oil, and reducing our greenhouse gas emissions."[45]
>
> —Barack Obama, the forty-fourth president of the United States.

As of 2013 the US federal government offered a $7,500 tax credit for purchases of electric vehicles. Some states also offer tax credits; California, for instance, offers a $5,000 credit. Many other countries, including Canada, Belgium, Spain, and the United Kingdom, similarly provide subsidies for purchases of electric cars. Government funds will provide the incentives that consumers need to try the new vehicles, while education programs promote their advantages. "Going back for generations, the basic process of buying and owning and fueling your vehicle has remained pretty much the same," explains Arun Banskota, the president of eVgo, a startup company that is building electric-car charging infrastructure. "We've learned how hard it is to change consumer behavior."[47] Providing a financial incentive is an integral part of encouraging consumers to take what they may perceive as a risk in buying an alternative energy vehicle.

Society Must Share in the Costs

The benefits of electric cars and other alternative energy vehicles go beyond the direct economic advantages accrued to manufacturers of the automobiles and components: Society as a whole benefits from having a cleaner environment. Other benefits result from having to rely less on oil, a nonrenewable resource. "The world's citizens and governments must accept that the Earth's resources are finite and commit themselves to the development of new power sources for automobiles,"[48] says Koji Omi, a member of the Japanese House of Representatives.

Because society benefits from alternative energy technologies, society should share the cost of developing and implementing them. Obama, who has committed federal funding to research and development of alternative energy vehicles, says of this commitment: "It's not just about saving money. It's also about saving the environment. But it's also about our national security. . . . It's not a Democratic idea or a Republican idea, it's just a smart idea."[49]

The internal combustion engine has dominated the world's highways and byways for more than a century and has had time to become firmly established as the technology of choice. Throughout the world, networks of gas stations support this technology. Alternative energy cars will not become competitive without government help.

Subsidies for automakers go back almost to when the first cars started rolling off the assembly line. "In the most insidious and unsuspected ways, the automobile industry has managed to line up subsidies for it-self," writes Francis de Winter, an engineering consultant who specializes in the energy sector. "These are hidden well enough so that many seem to be unaware that the automobile is a very costly luxury for society."[50]

The government also has heavily sub-sidized the development of fossil fuels. According to Oil Change International, the United States provides between $10 billion and $52 billion to the oil, gas, and coal industries. "Beneath these sub-sidies lie a host of other expenses that also support the oil, gas, and coal indus-tries," writes Oil Change International. "States provide billions more in sub-sidies."[51] The fossil fuel industry is, by now, well established. It no longer needs government subsidies. Rather than subsidizing the current technologies, the government should direct these funds to new technologies that will support a better future.

> "The world's citizens and governments must accept that the Earth's resources are finite and commit themselves to the development of new power sources for automobiles."[48]
>
> —Koji Omi, member of Japan's House of Representatives.

The Government Should Not Be Involved in Developing Alternative Energy Cars

"No evidence exists to suggest that the government has better knowledge to make investment decisions or to commercialize technologies when the private sector chooses to bypass these opportunities. If there is a role for alternative sources in America's energy portfolio, it should be dictated by price and competition, not government handouts."

—Nicolas Loris, policy analyst at the Heritage Foundation, a conservative think tank.

Quoted in *U.S. News & World Report*, "Competition, Not Handouts, Should Determine Role of Green Energy," January 18, 2012. www.usnews.com.

Government involvement in the automobile industry has consistently failed to make a difference in the viability of alternative energy vehicles. Obama promoted electric vehicles as a component of his green initiative and set a goal to put 1 million electric cars on the road by 2015. By January 2013 Obama had reversed course, realizing that his plan simply was not viable. Jerry Taylor, a senior fellow at the Cato Institute, writes, "The Obama administration has spent $5 billion to promote the manufacture of electric vehicles so as to put 1 million EVs on American roads by 2015. But layoffs and bankruptcies have plagued those receiving EV handouts because the technology is still problematic and the final product so expensive that consumers won't buy it, even with $7,500 rebates."[52]

In automobile manufacturing, as in any other industry, experience shows that free market economics work best. In a free market only the technologies that are economically viable will succeed. Consumer preferences and the private sector should dictate auto manufacturing. And yet the US government has invested billions of dollars to develop alternative energy vehicles that consumers seemingly do not want at any price. For instance, the 2009 stimulus bill included $2 billion for the development and manufacture of electric-car batteries and other components, and the Energy Department has provided up to $25 billion in direct federal loans as part of its advanced-technology vehicle program. Meanwhile, the government has spent another $400,000 on education programs to get Americans to buy the cars.

> "Layoffs and bankruptcies have plagued those receiving EV handouts because the technology is still problematic and the final product so expensive that consumers won't buy it, even with $7,500 rebates."[52]
>
> —Jerry Taylor, a senior fellow at the Cato Institute.

Almost five years later, few electric cars are actually being sold. What the government has really accomplished is a larger supply of these vehicles than can be absorbed by the demand. Chevrolet twice stopped production of the Volt in 2012 because they were not moving from dealer lots, despite a significant drop in price. In a 2012 audit of government spending on alternative energy vehicles, the Congressional Budget Office concluded that the sales of electric-drive vehicles are so underwhelming that the billions of government dollars invested will have "little or no impact on total gasoline consumption and greenhouse-gas emissions."[53]

Mandates and Quotas

In addition to grants, loans, and other financial incentives, governments have attempted to use quotas to require automakers to introduce alternative energy vehicles. The failure of such government mandates is evident in the fact that the government consistently backs off its lofty goals. Take, for instance, California's policies. California's Zero Emission Vehicle (ZEV) Mandate, which required 2 percent of the state's vehicles to have no emissions by 1998 and 10 percent by 2003, proved ineffectual—automobile makers simply could not comply. As a result, the state weakened its policies repeatedly over the next decade, ultimately resulting in automakers' recall and destruction of the electric vehicles that they had produced to comply with the mandate. The state of California has also established quotas mandating that a certain percentage of cars made by any given automaker should be zero-emission vehicles. Automobile manufacturers that do not meet these quotas can buy credits from those that are over their limit. To date, this has resulted not in more automakers with more alternative energy models on the market, but rather in a shift of wealth from one automaker to another. According to a Morgan Stanley report, Tesla made $40.5 million on credits in 2012 and $67.9 million in the first quarter of 2013.

The Rich Get Richer

Some of the money that government is investing goes directly to consumers. The cars are more expensive than comparable conventional vehicles; to get consumers to buy them, governments offer bribes in the form of subsidies. In the United States, purchasers receive up to $7,500 in a tax rebate from the federal government and thousands more from some state governments. The United States is not the only country subsidizing the purchase of alternative fuel vehicles: many countries in Europe and other parts of the world also provide these incentives.

One of the criticisms of these subsidies is that they go to wealthy Americans who can afford cars like the $100,000 Tesla Model S—wealthy Americans who are least in need of government funds. In addition, Tesla also received a $465 million loan guarantee and a $10 million grant from the California Energy Commission. "Taxpayers pay first so Tesla can build the cars and again to help the wealthy buy them," writes the author of a *Wall Street Journal* editorial. "Why should middle-class taxpayers whose incomes are falling still pay to subsidize the purchase of cars that only the affluent can afford?"[54]

Undermining Market Forces

Still, Tesla is one of the few examples of success. Most federal subsidies, loans, and grants have done little to help companies focused on alternative energy vehicles and/or the fuels that power them to become economically viable. In 1993 the DOE and the Big Three US automakers—Ford, General Motors, and DaimlerChrysler—forged the Partnerships for a New Generation of Vehicles (PNGV) program to build production prototypes with a threefold improvement in fuel economy. More than $250 million per year was directed to the cause. Each of the Big Three automakers came out with prototypes for a hybrid electric in 2000, but none of these was put into production. Meanwhile, Toyota, inspired and/or driven by market forces rather than government handouts, unveiled its hybrid in Japan in 1998 and then in the United States in 2000.

> "While [tax] credits have put millions of FFVs on the road since the late 1990s, most have been large SUVs, pickups, and sedans that get relatively poor gas mileage."[56]
>
> —*Consumer Reports,* a monthly magazine that tests and publishes reviews of consumer products and services.

Several of the other automobile companies and battery makers that have received taxpayer funds have floundered. Fisker Automotive, an American automaker based in Anaheim, California, received hundreds of millions of dollars in loans and grants to produce the Fisker Karma, a plug-in hybrid luxury sports sedan that debuted at the 2008 North American International Auto Show. Despite glowing reviews, Fisker could translate neither the Karma nor subsequent all-electric concept cars into viable options for consumers. In 2013 Fisker laid off three-quarters of its workforce and announced that it was preparing for bankruptcy, negating any potential benefit from the taxpayers' investment.

Among the problems that Fisker cited was the earlier bankruptcy of A123 Systems, the maker of its lithium-ion batteries. A123 Systems, which had received a $249 million loan from the US government, began bankruptcy proceedings in 2012. Another auto battery maker, Ener1, Inc., which received a $118.5 million taxpayer-funded grant, also went bankrupt. An editorial in the *Washington Times* concludes that the federal support "accomplishes nothing beyond propping up—for a time— some fly-by-night companies."[55]

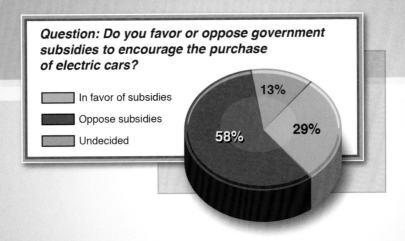

Majority Does Not Want to Subsidize Electric Car Buyers

The federal government currently provides up to $7,500 in tax credits to purchasers of electric vehicles. Commenting that this has not done enough to spur sales, President Obama has proposed increasing the subsidy to $10,000 a car. As shown in this chart illustrating the opinions of Americans from a February 2012 poll, most Americans do not favor the taxpayer-funded subsidies.

Question: Do you favor or oppose government subsidies to encourage the purchase of electric cars?

- In favor of subsidies
- Oppose subsidies
- Undecided

13%

29%

58%

Source: Rasmussen Reports, "Trickle Down Environmentalism Has Little Public Support," February 24, 2012. www.rasmussenreports.com.

Too Many Loopholes

Equally troubling are loopholes that allow automakers to benefit from government funding without making a true change. Flex-fuel vehicles (FFVs), which can use blends of ethanol up to 85 percent, are an example. Manufacturers of FFVs currently receive incentives in the form of fuel economy credits earned for compliance with fuel economy standards set by the federal government (known as Corporate Average Fuel Economy, or CAFE). "While the credits have put millions of FFVs on the road since the late 1990s," says *Consumer Reports*, "most have been large SUVs, pickups, and sedans that get relatively poor gas mileage and don't do well in *Consumer Reports* testing."[56]

The flex-fuel debate highlights what happens when the government uses funding to try to dictate consumer behavior. Many owners of flex-fuel vehicles are not even aware that their car has a fuel-flex option, and most of those who are aware that there is a fuel option still chose to use regular gasoline instead. Gasoline not only gets better gas mileage, it is also more convenient. In fact, only 1 percent of gas stations in the United States offer ethanol. Outside the Midwest it can be difficult to find and, despite government subsidies to bring the price in line with gasoline, is often still more expensive. "Many of today's vehicles are flex-fuel compatible," said Dennis Virag, president of the Automotive Consulting Group Inc. "The problem is that relatively few people are using flex fuels, given the price and availability of gasoline. You can't force it on the consumer unless you ban gasoline, which will never happen."[57] With the vast majority of Americans using gasoline in their fuel-flex vehicles, the public investment in fuel-flex technologies—like many of the other alternative energy options—is a true waste of taxpayer dollars.

Source Notes

Overview: Visions of the Future: Alternative Energy Cars

1. Franz von Holzhausen, "Tesla Model S Features & Specs." www.teslamotors.com.
2. Quoted in Lydia Bjornlund, "Green with Envy: The Tesla Model S," *Innovation*, Fall 2013.

Chapter One: Are Alternative Energy Cars Affordable?

3. Quoted in *Guardian*, "Will Electric Cars Ever Enter the Mainstream?," July 3, 2013. www.rawstory.com.
4. Quoted in Brad Tuttle, "What Would Make an All-Electric Car Appeal to the Masses?," *Time*, January 13, 2013. http://business.time.com.
5. US Department of Energy, "Hydrogen." www.fueleconomy.gov.
6. John Peterson, "Electric Vehicles: The Opportunity of Which Decade?," Alt Energy Stocks, January 27, 2011. www.altenergystocks.com.
7. John Hoeven, "Why We Need the Keystone Oil Pipeline," CNN Opinion, February 24, 2012. www.cnn.com.
8. NADA, "Plug-In Electric Vehicles: Market Analysis and Used Price Forecast." www.nada.com.
9. Quoted in Dave Destler, "Tesla Roadster: Performance, Style and Efficiency in a Zero-Emissions Package," *Innovation*, Spring 2013, p. 21.
10. Quoted in Susanna Schick, "The Future of Affordable EVs According to John Viera," Sustainable Brands 2013, June 10, 2013. http://gas2.org.
11. Quoted in Jeff Bennett, "Volt Falls to Electric-Car Price War," *Wall Street Journal*, August 6, 2013, p. B2.
12. Quoted in Clifford Krauss, "Sudden Spike in Gas Prices, but Increases May Be Short-Lived," *New York Times*, July 12, 2013. www.nytimes.com.

13. Jason Steele, "Hybrid Cars Pros and Cons—Benefits & Problems," (blog), *Money Crashers*. www.moneycrashers.com.

14. Thomas J. Knipe, Loic Gaillac, and Juan Argueta, "100,000-Mile Evaluation of the Toyota RAV4 EV," EV Charger News. www.ev chargernews.com.

Chapter Two: Are Alternative Energy Cars a Viable Option?

15. Chris Paine, "Five Myths About Electric Cars," *Washington Post*, April 26, 2013. http://articles.washingtonpost.com.

16. Quoted in Paine, "Five Myths About Electric Cars."

17. Quoted in Ryan Bradley, "Elon Musk's Moment," *CNN Money*, May 30, 2013. http://tech.fortune.cnn.com.

18. John Voelcker, "Toyota Hydrogen Fuel-Cell Car to Come in 2015, for $50,000," *Green Car Reports*, May 10, 2010. www.greencarre ports.com.

19. Etim U. Ubong, "From Internal Combustion Engine to Hybrid Propulsion," *Advances in Automobile Engineering*, April 27, 2012. www .omicsgroup.org.

20. Quoted in Daniel Sperling and Deborah Gordon, *Two Billion Cars: Driving Toward Sustainability*. New York: Oxford University Press, 2009, p. 106.

21. Green Fuel Online, "Hydrogen," 2013. www.greenfuelonline.com.

22. Toyota, *Toyota FCHB Book*, p. 6. www.toyotaaruba.com.

23. Quoted in Brad Plumer, "What's Wrong with the Electric Car? Psychology, Perhaps," *Washington Post Wonkblog*, November 23, 2011. www.washingtonpost.com.

24. Quoted in Thomas Tamblyn, "Ultra Long-Range Electric Cars Possible but Not Viable Says Nissan Engineer," T3: The Gadget Website, July 3, 2013. www.t3.com.

25. Louis Woodhill, "Electric Cars Are an Extraordinarily Bad Idea," *Forbes*, September 14, 2011. www.forbes.com.

26. Norihiko Shirouzu, Yoko Kubota, and Paul Lienert, "Are Electric Cars Running Out of Juice Again?," *Reuters*, February 4, 2013. www .reuters.com.

27. Patrick Michaels, "If Tesla Would Stop Selling Cars, We'd All Save Some Money," *Forbes*, May 27, 2013. www.forbes.com.

28. Quoted in Tuttle, "What Would Make an All-Electric Car Appeal to the Masses?"

29. Quoted in John Voelcker, "Hydrogen Fuel-Cell Cars Not Viable, Says Volkswagen CEO," *Green Car Reports*, March 21, 2013. www.greencarreports.com.

30. Quoted in National Research Council, *Transitions to Alternative Vehicles and Fuels*. Washington, DC: National Academies Press, 2013, p. vii.

Chapter Three: Do Alternative Energy Cars Benefit the Environment?

31. Troy R. Hawkins, et al., "Comparative Environmental Life Cycle Assessment of Conventional and Electric Vehicles," *Journal of Industrial Ecology*, October 4, 2012, p. 53. http://onlinelibrary.wiley.com.

32. Quoted in Paine, "Five Myths About Electric Cars."

33. Don Anair, "Future State of Charge: How Clean Will Electric Vehicles Get?," *Equation* (blog), August 13, 2012. http://blog.ucsusa.org.

34. Union of Concerned Scientists, "Fueling a Better Future: The Many Benefits of 'Half the Oil.'" www.ucsusa.org.

35. National Wildlife Federation, *Restoring a Degraded Gulf of Mexico: Wildlife and Wetlands Three Years into the Gulf Oil Disaster*, April 2013. www.nwf.org.

36. James Conca, "Are Electric Cars Really That Polluting?," *Forbes*, July 21, 2013. www.forbes.com.

37. Quoted in James R. Healey, "Electric-Car Costs Can Outweigh Cheap Fuel," *USA Today*, June 12, 2013. www.usatoday.com.

38. Bjørn Lomborg, "Electric Cars Will Be Great Someday. Just Not Now," *Slate*, April 14, 2013. www.slate.com.

39. Ozzie Zehner, "Unclean at Any Speed: Electric Cars Don't Solve the Automobile's Environmental Problems," *IEEE Spectrum*, June 30, 2013. http://spectrum.ieee.org.

40. Brigham McCown, "Electric Cars and Their Dirty Secret," *National Journal Energy Experts Blog*, June 14, 2013. http://energy.national journal.com.

41. Quoted in Mike Ives, "Boom in Mining Rare Earths Poses Mounting Toxic Risks," *Environment 360*, January 28, 2013. http://e360.yale.edu.

42. Quoted in Lawrence Ulrich, "Boom Times for Born-Again Diesels," *New York Times*, July 21, 2013, p. C10.

43. Troy R. Hawkins, Ola Moa Gausen, and Anders Hammer Strømman, "Environmental Impacts of Hybrid and Electric Vehicles: A Review," *International Journal of Life Cycle Assessment*, September 2012, p. 998.

Chapter Four: Should the Government Support the Development of Alternative Energy Cars?

44. Quoted in Amy Harder, "What's the Future of Electric Cars?," *National Journal Energy Experts Blog*, June 17, 2013. http://energy.nationaljournal.com.

45. US Department of Energy Loan Programs Office, "Obama Administration Awards First Three Auto Loans for Advanced Technologies," June 23, 2009.

46. National Research Council, *Transitions to Alternative Vehicles and Fuels*, p. 9.

47. Quoted in *Guardian*, "Will Electric Cars Ever Enter the Mainstream?"

48. Koji Omi, "Alternative Energy for Transportation, *Issues in Science and Technology*, Summer 2009. www.issues.org.

49. Quoted in Securing America's Future Energy, "President Obama Discusses SAFE and the Energy Security Trust at Argonne National Labs," March 15, 2013. www.secureenergy.org.

50. Francis de Winter, "Insidious Subsidies Granted to the Automobile," Ecotopia. www.ecotopia.com.

51. Oil Change International, "Capitol Spill: How Congress Leaks Favors to Fossil Fuels," Price of Oil. http://priceofoil.org.

52. Jerry Taylor, "It's Up to the Private Sector to Invest in New Technology," *U.S. News & World Report*, January 18, 2012. www.usnews.com.

53. Quoted in *Washington Times*, "Editorial: Unplug the Electric Subsidies," March 18, 2013. www.washingtontimes.com.

54. *Wall Street Journal*, "The Other Government Motors," editorial, May 23, 2013. http://online.wsj.com.

55. *Washington Times*, "Editorial: Unplug the Electric Subsidies," March 18, 2013. www.washingtontimes.com.

56. *Consumer Reports*, "The Great Ethanol Debate," January 2011. www.consumerreports.org.

57. Quoted in Louis Jacobson, "Flex-Fuel Vehicles Are Nowhere Near Universal," PolitiFact.com, *Tampa Bay Times*, February 8, 2012. www.politifact.com.

Alternative Energy Car Facts

Components of Alternative Energy Cars

- The battery is by far the heaviest and most expensive part of electric vehicles, weighing more than 1,000 pounds and costing upwards of $10,000.
- An electric car has far fewer parts than an internal combustion engine. Compared to the hundreds of parts needed to power an internal combustion engine, an electric vehicle has just four main parts: a potentiometer, batteries, a direct current controller, and a motor.
- Electric motors are inherently more efficient than combustion engines, effectively utilizing at the wheels more than 90 percent of the energy provided, compared to 37 percent for today's conventional car engine.

Sales and Industry Growth

- In 2012 there were 2.3 million registered hybrid models on American roads, according to R.L. Polk registration data.
- The Electric Drive Transportation Association reports that the market share of hybrid and pure electric cars in 2012 was 3.38 percent.
- According to the National Research Council, hybrid vehicles (in 2013) are the fastest growing segment of the light-duty vehicle market, which includes cars, SUVs, minivans, and pickup trucks.
- As of January 2013 there were a total of about thirty thousand electric vehicles on American roads.
- According to Green Car Congress, hybrid sales in Japan have rocketed to 40 percent of the industry total, with the Prius a top seller. In the United States, the Prius is the third best-selling automobile.

Consumer Acceptance

- In the Global Auto Executive Survey of 2013, 62 percent of automotive executives said that the use of alternative fuel technologies was "extremely" or "very" important to consumers.
- According to a 2013 report by the professional services company KPMG, 36 percent of automotive executives believe plug-in hybrids will attract most consumer demand, and 24 percent of automotive executives are considering big investments in plug-in hybrids.
- In a Consumer Affairs poll undertaken in June 2012, 73 percent of drivers said they would consider an alternatively fueled vehicle.
- A USA Today/Gallup poll found that 57 percent of Americans say they will not buy an electric car because of concerns about range.
- In a 2012 survey conducted by J.D. Power and Associates, 94 percent of US drivers said their next car would *not* be a plug-in.
- Only 17 percent of respondents to the 2013 Global Auto Executive Survey believe that fuel cell technologies have the potential for significant sales by 2018. In China, however, 44 percent ranked fuel cell technologies as the most popular emerging technology.

Fuel Efficiency

- The Environmental Protection Agency found that the 2012 domestic fleet was the cleanest and most fuel-efficient fleet ever, with an overall average of 23.8 miles per gallon.
- Statistics from the US Energy Information Administration (EIA) show that 60 percent of transportation fuel consumption is gasoline used by motor vehicles. The EIA predicts that this will decline to 47 percent by 2040, largely due to alternative fuels.
- Fuel economy of hybrids is at least 25 percent better than conventional gasoline-powered vehicles.
- The DOE has calculated that a typical electric vehicle can run for 43 miles on a dollar's worth of electricity.

Government Funding and Programs

- In 1993 the DOE initiated a hybrid electric vehicle (HEV) program as a cost-sharing partnership with General Motors, Ford, and

DaimlerChrysler in which the three automakers committed to producing market-ready HEVs by 2003.

- According to a KPMG 2013 report, 69 percent of automotive executives said that government subsidies are needed for electric vehicles to become affordable.
- In September 2013 the Congressional Budget Office reported that federal policies to prop up and promote electric cars will cost taxpayers $7.5 billion through 2019.

Related Organizations and Websites

Alliance of Automobile Manufacturers
803 Seventh St. NW, Suite 300
Washington, DC 20001
phone: (202) 326-5500
website: www.autoalliance.org

The Alliance of Automobile Manufacturers is an association with twelve global vehicle manufacturers that serves as the leading advocacy group for the auto industry.

California Cars Initiative
323 Los Altos Dr.
Aptos, CA 95003
phone: (916) 371-2800
e-mail: info@calcars.org
website: www.calcars.org

CalCars is a Palo Alto–based nonprofit startup of entrepreneurs, engineers, environmentalists, and others focused on promoting plug-in hybrid electric vehicles through public policy, technology development, and promotion to consumers.

California Fuel Cell Partnership (CaFCP)
3300 Industrial Blvd., Suite 1000
West Sacramento, CA 95691
phone: (916) 371-2870
e-mail: info@cafcp.org
website: www.cafcp.org

The CaFCP is committed to promoting fuel cell vehicles as a means of moving toward a sustainable energy future, increasing energy efficiency, and reducing or eliminating air pollution and greenhouse gas emissions. The group publishes press releases, newsletters, and information sheets concerning hydrogen power and fuel cell technology.

Electric Auto Association
323 Los Altos Dr.
Aptos, CA 95003
phone: (831) 688-8669
website: www.electricauto.org

The Electric Auto Association (EAA) is a nonprofit educational organization formed in 1967 by Walter Laski in San Jose, California. The goal of the EAA is to promote the advancement and widespread adoption of electric vehicles.

Electric Drive Transportation Association (EDTA)
1250 Eye St. NW, Suite 902
Washington, DC 20005
phone: (202) 408-0774
website: www.electricdrive.org

The EDTA conducts public policy, advocacy, education, and research programs to promote electric drive as the best means to achieve a clean source of energy in the transportation industry.

Independent Petroleum Association of America (IPAA)
1201 Fifteenth St. NW, Suite 300
Washington, DC 20005
phone: (202) 857-4722
fax: (202) 857-4799
website: www.ipaa.org

The IPAA represents thousands of independent crude oil and natural gas explorer/producers in the United States and provides advocacy, public policy, and education support for the industry.

National Fuel Cell Research Center (NFCRC)
University of California
Irvine, CA 92697
phone: (949) 824-1999
fax: (949) 824-7423

The NFCRC was founded to provide leadership in the preparation of educational materials and programs throughout the country. The NFCRC engages undergraduate and graduate students from all disciplines of engineering and the physical and biological sciences and collaborates on courses and team projects.

National Renewable Energy Laboratory (NREL)
15013 Denver West Pkwy.
Golden, CO 80401
phone: (303) 275-3000
website: www.nrel.gov

NREL is a federal laboratory dedicated to the research, development, commercialization, and deployment of renewable energy and energy efficiency programs.

Union of Concerned Scientists
2 Brattle Square
Cambridge, MA 02138
phone: (617) 547-5552
e-mail: www.ucsusa.org

The Union of Concerned Scientists is an independent nonprofit alliance of more than four hundred thousand citizens and scientists who focus on scientific analysis to protect the environment and conserve natural resources.

US Department of Energy
Energy Efficiency & Renewable Energy
Vehicle Technologies Office
1000 Independence Ave. SW
Washington, DC 20585
phone: (202) 586-5000
website: www1.eere.energy.gov

This federal program was created to develop and support more energy efficient and environmentally friendly transportation technologies that will reduce the use of petroleum.

For Further Research

Books

James Billmaier, *Jolt! The Impending Dominance of the Electric Car and Why America Must Take Charge*. Charleston, SC: Advantage, 2010.

Lynn J. Cunningham et al., *Alternative Fuel and Advanced Vehicle Technology Incentives: A Summary of Federal Programs*. Washington, DC: Congressional Research Service, 2012.

James D. Halderman, *Hybrid and Alternative Fuel Vehicles*. New Jersey: Prentice Hall, 2012.

Iqbal Husain, *Electric and Hybrid Vehicles: Design Fundamentals*. 2nd ed. Boca Raton, FL: CRC, 2010.

Jim Motavalli, *High Voltage: The Fast Track to Plug In the Auto Industry*. New York: Rodale, 2011.

National Research Council, *Transitions to Alternative Vehicles and Fuels*. Washington, DC: National Academies Press, 2013.

Daniel Sperling and Deborah Gordon, *Two Billion Cars: Driving Toward Sustainability*. New York: Oxford University Press, 2009.

Periodicals

Consumer Reports, "Pros and Cons: A Reality Check on Alternative Fuels," May 2011.

Nichola Groom, "The 'California Hydrogen Highway' Failure and Other Struggles in the Electric Vehicle Market," *Huffington Post*, May 20, 2013.

Drew Hendricks, "3 Ways Alternative Energy Cars Can Save Drivers Money in 2013," *Huffington Post*, March 8, 2013.

Bjørn Lomborg, "Electric Cars Will Be Great Someday. Just Not Now," *Slate*, April 14, 2013.

Brad Tuttle, "What Would Make an All-Electric Car Appeal to the Masses?," *Time*, January 29, 2013.

Wall Street Journal, "The Other Government Motors: Tesla by the Numbers: How Taxpayers Made an Electric Car Company," May 23, 2013.

Internet Sources

Department of Energy, Energy Efficiency & Renewable Energy, "Vehicle Technologies Office." www1.eere.energy.gov/vehiclesandfuels.

KPMG, "2013 Global Automotive Executive Survey." www.kpmg.com /global/en/issuesandinsights/articlespublications/global-automotive -executive-survey/pages/default.aspx.

PBS, "Timeline: History of the Electric Car." www.pbs.org/now/shows /223/electric-car-timeline.html.

Union of Concerned Scientists, "State of Charge: Electric Vehicles' Global Warming Emissions and Fuel-Cost Savings Across the United States," June 2012. www.ucsusa.org/clean_vehicles/smart-transportation -solutions/advanced-vehicle-technologies/electric-cars/emissions-and -charging-costs-electric-cars.html.

US Bureau of Labor Statistics, "Careers in Electric Vehicles." www.bls .gov/green/electric_vehicles/#chart1.

US Energy Information Administration, "Annual Energy Outlook 2013: With Projections to 2040," April 2013. www.eia.gov/forecasts/aeo /pdf/0383(2013).pdf.

Who Killed the Electric Car?, documentary, Papercut Films, 2007. www .whokilledtheelectriccar.com/takeaction.

Index

About the Author

Lydia Bjornlund is a freelance writer and editor living in northern Virginia. She has written more than two dozen nonfiction books for children and teens, mostly on American history and health-related topics. She also writes books and training materials for adults on issues related to land conservation, emergency management, public policy, and industrial design—including cool new products like electric cars. Bjornlund holds a master's degree in education from Harvard University and a BA in American Studies from Williams College. She lives with her husband, Gerry Hoetmer, and their children, Jake and Sophia.